I0506578

Brainy Puzzler Group

CONTENTS

HOW TO PLAY TIC TAC TOE

• This is a game for two players.

• One player chooses X and one chooses O.

• Each player takes it in turns to place their mark (X or O) in the 3x3 grid.

• The winner is the first player to get three in a row horizontally, vertically or diagonally.

• The game is tied if the board is filled without either player scoring a full row of three.

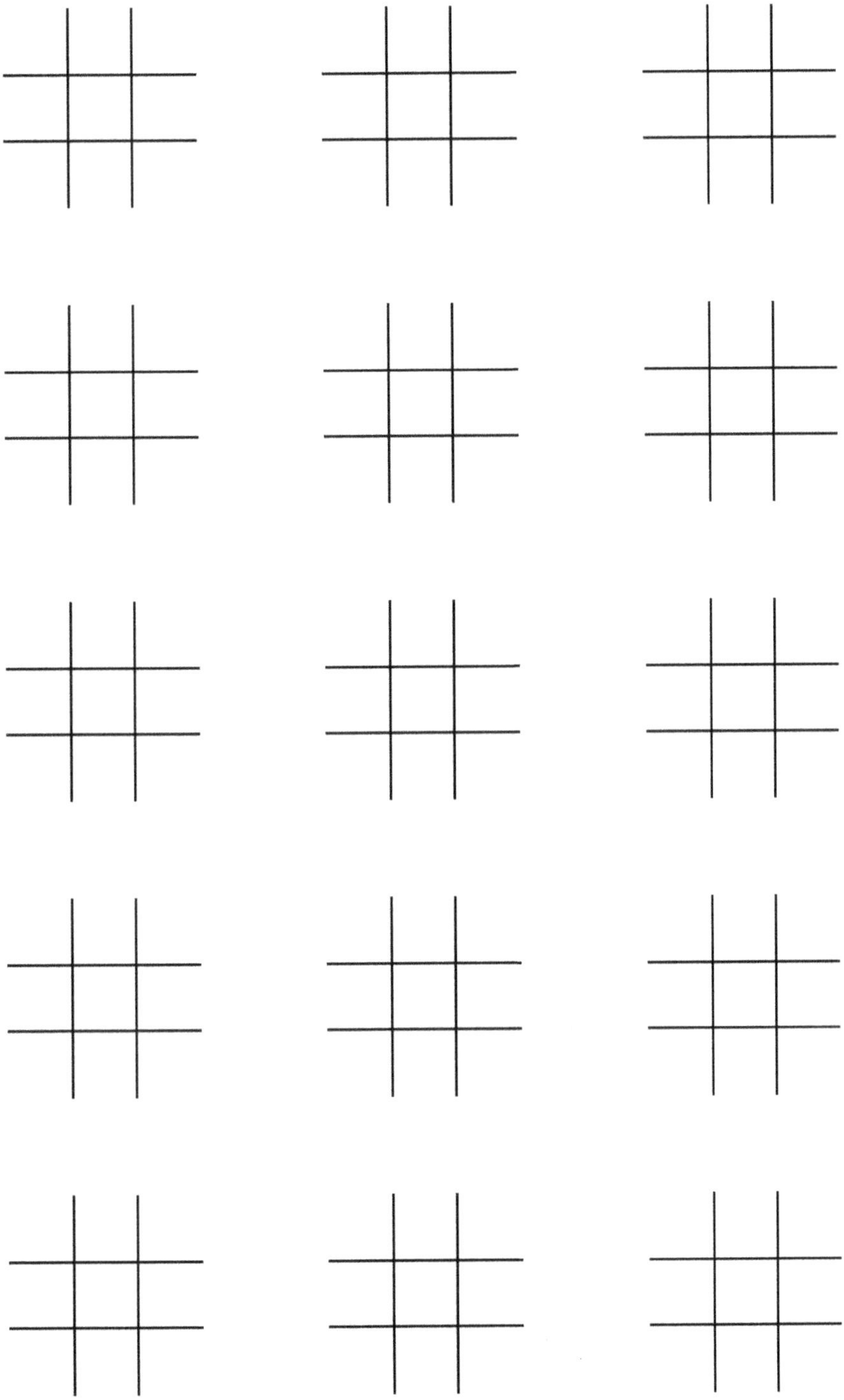

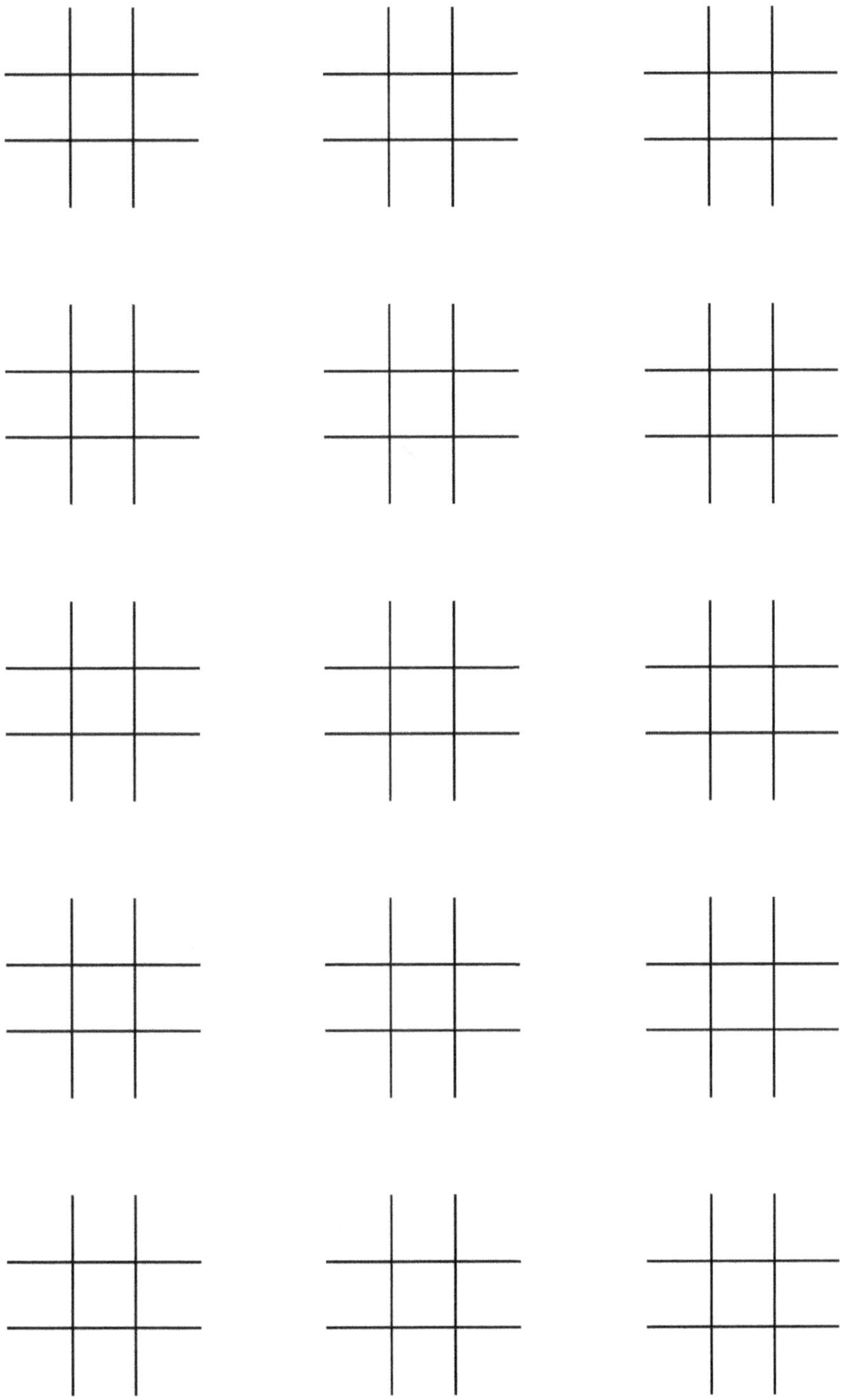

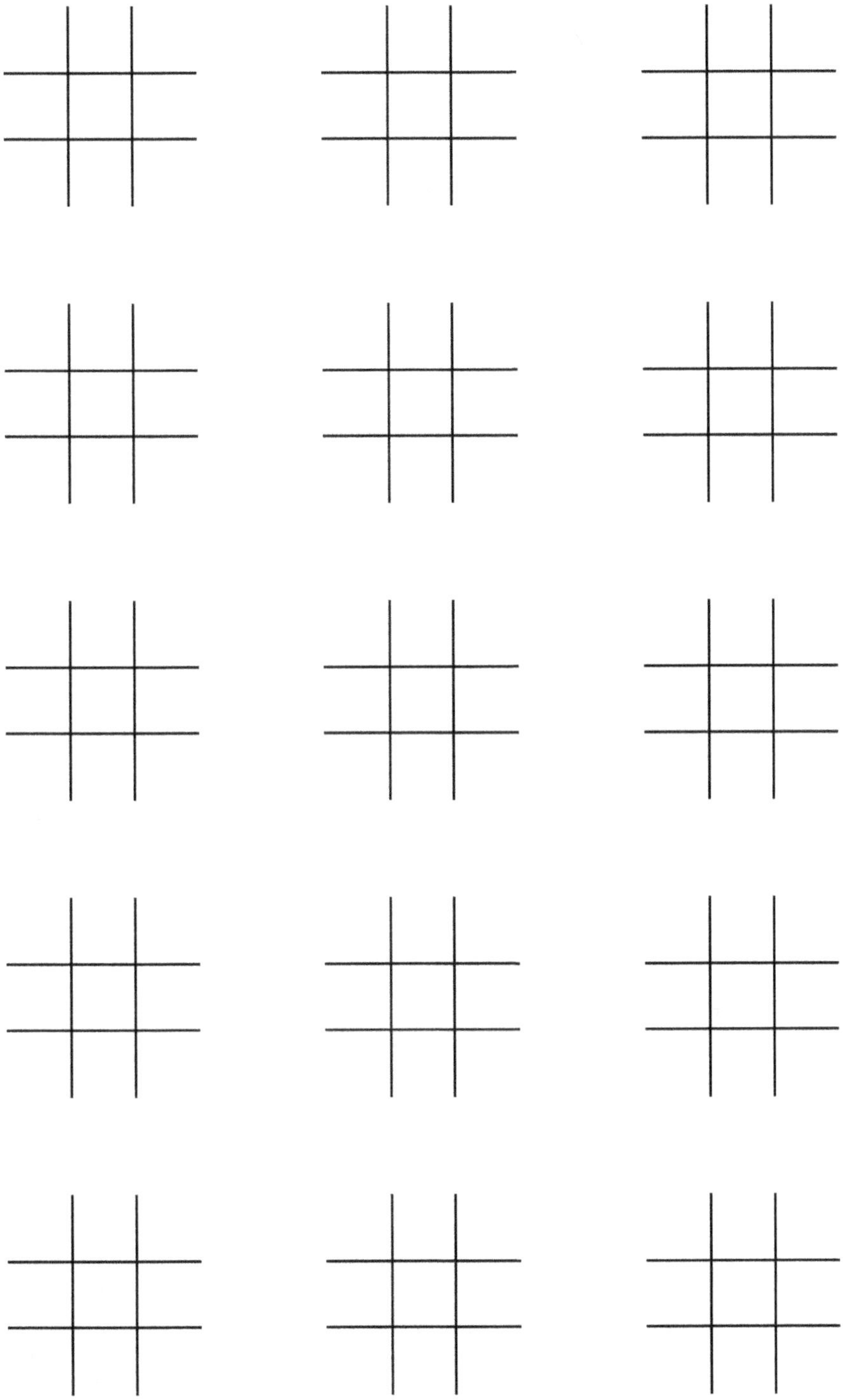

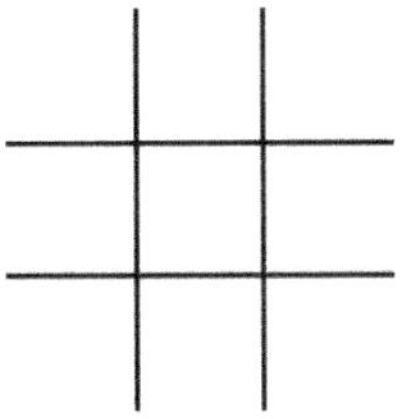
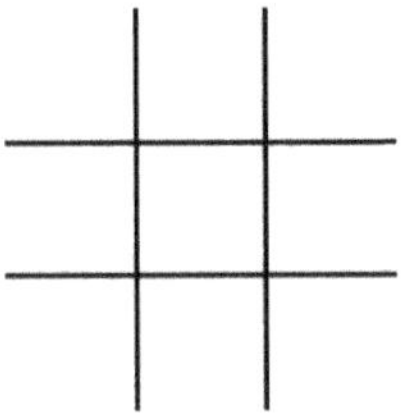
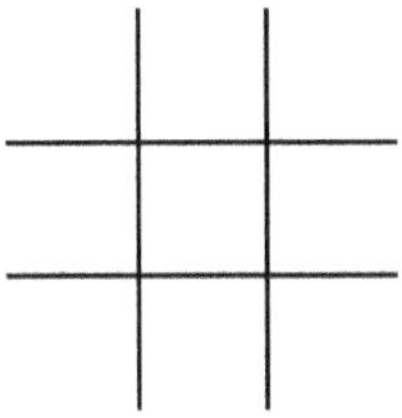
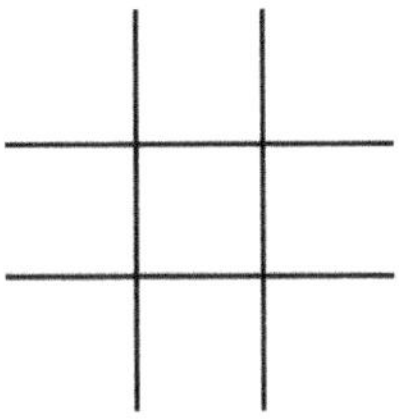
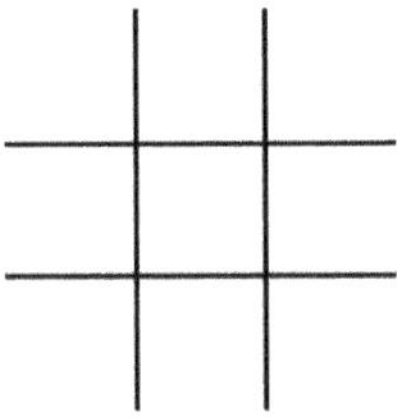
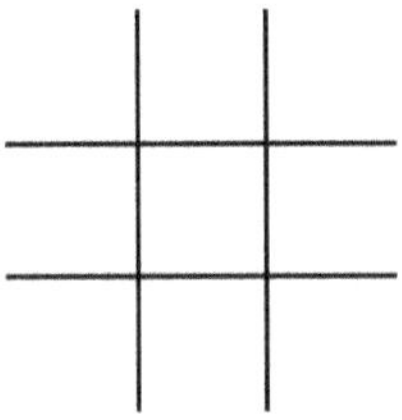
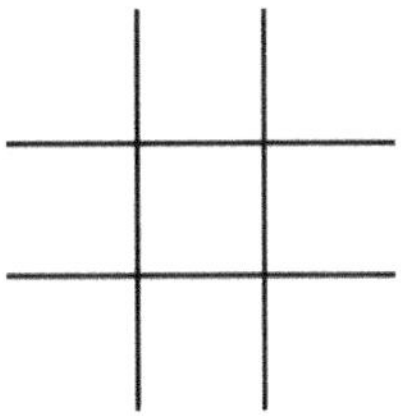
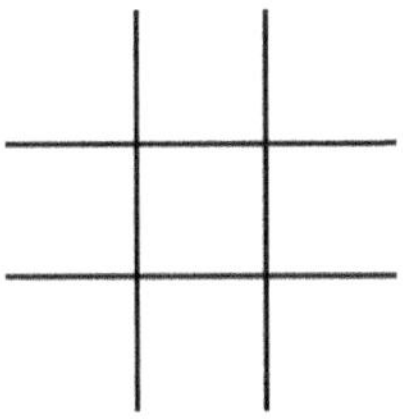
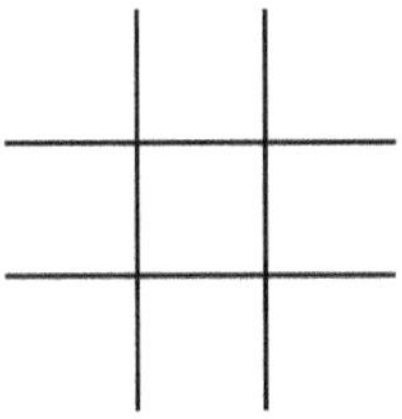
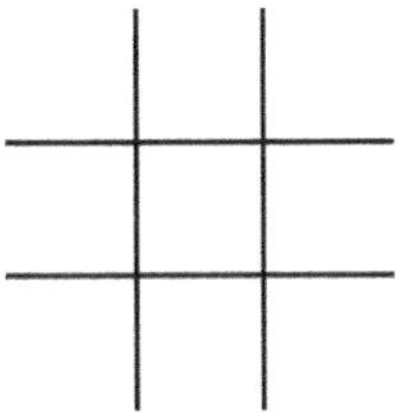
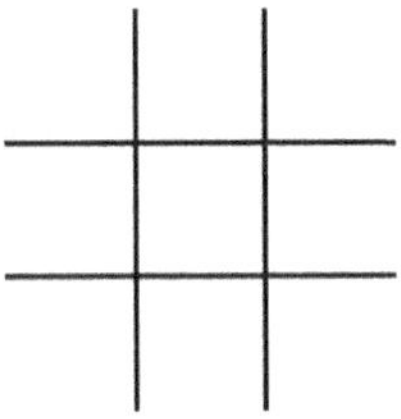
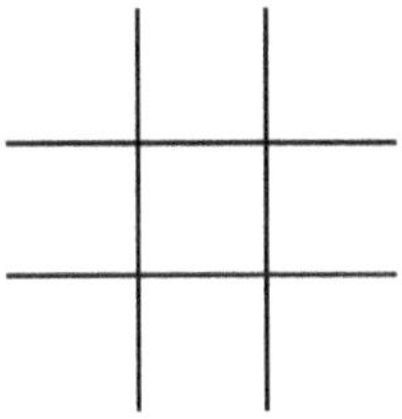
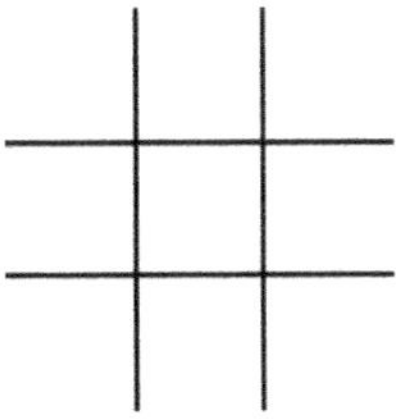
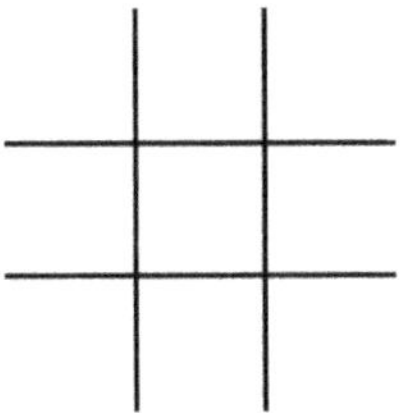
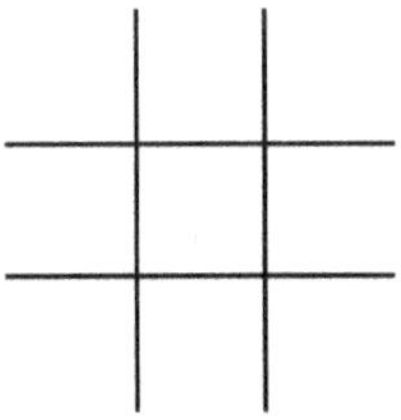

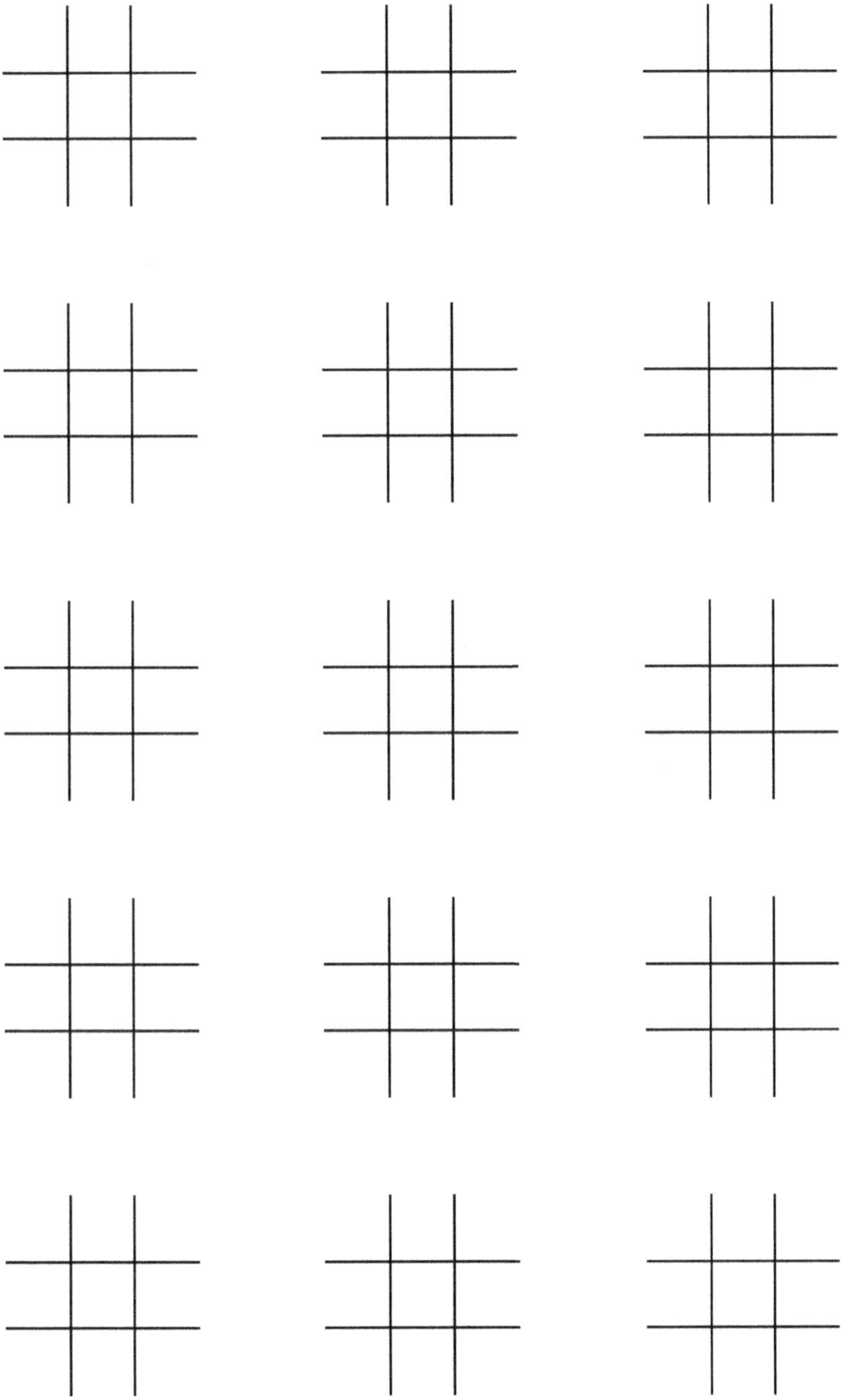

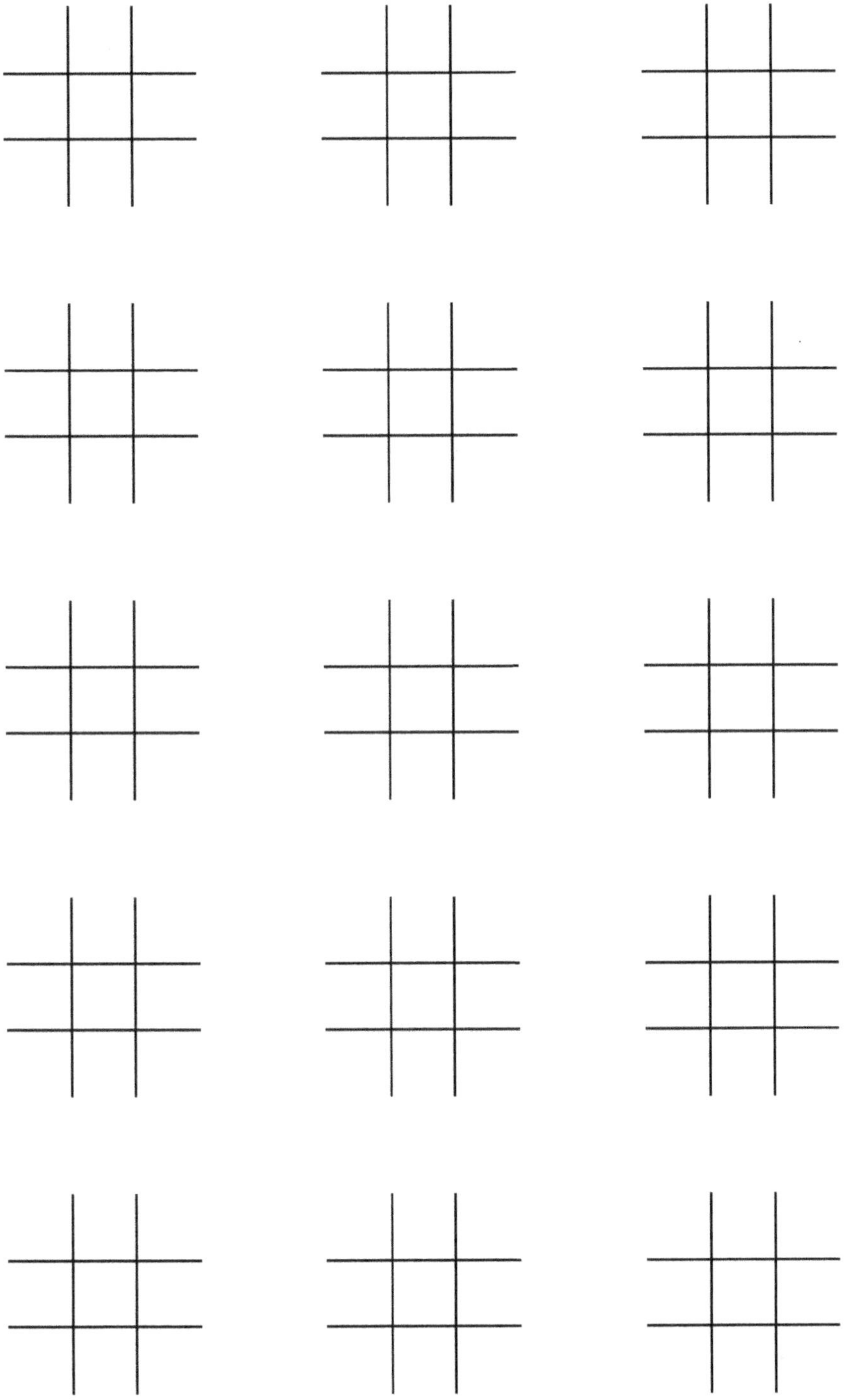

HOW TO PLAY SEA BATTLE

• To play Sea Battle, you and your opponent will each need one page of the book.

• Before play begins, secretly arrange your 'ships' on your board by drawing rectangles around the squares they occupy. An aircraft carrier for example will be one rectangle that surrounds five squares in a row. Each player has one aircraft carrier, one battleship, one cruiser, two destroyers and three submarines (which each occupy just one square each).

• Once both players have drawn their ships on the board, play begins.

• The first player will call a grid square, for example, "B3"

• Their opponent will answer 'Hit' or 'Miss.'

• If they 'hit' their opponent's ship, they mark that square with a circle on the opponent board and make another guess. Guesses continue until they 'miss.' Misses should be marked with a cross on the opponent board.

• Once the player misses, it is the opponent's turn to guess.

• Play continues until one player sinks all of his opponents ships.

• Salvo variation: Advanced players may fire an agreed amount of multiple shots (guesses) at a time and may answer like this, "Three hits and two misses" so their opponent must deduce which shots were successful.

Me

	1	2	3	4	5	6	7	8	9	10
A										
B										
C										
D										
E										
F										
G										
H										
I										
J										

Aircraft Carrier | A | A | A | A | A |

Frigate | F | F | F | F |

Cruiser | C | C | C |

Destroyers | D | D | | D | D |

Submarines | S | | S | | S |

Opponent

	1	2	3	4	5	6	7	8	9	10
A										
B										
C										
D										
E										
F										
G										
H										
I										
J										

Aircraft Carrier | A | A | A | A | A |

Frigate | F | F | F | F |

Cruiser | C | C | C |

Destroyers | D | D | | D | D |

Submarines | S | | S | | S |

Me

	1	2	3	4	5	6	7	8	9	10
A										
B										
C										
D										
E										
F										
G										
H										
I										
J										

Aircraft Carrier | A | A | A | A | A |

Frigate | F | F | F | F |

Cruiser | C | C | C |

Destroyers | D | D | | D | D |

Submarines | S | | S | | S |

Opponent

	1	2	3	4	5	6	7	8	9	10
A										
B										
C										
D										
E										
F										
G										
H										
I										
J										

Aircraft Carrier | A | A | A | A | A |

Frigate | F | F | F | F |

Cruiser | C | C | C |

Destroyers | D | D | | D | D |

Submarines | S | | S | | S |

Me

	1	2	3	4	5	6	7	8	9	10
A										
B										
C										
D										
E										
F										
G										
H										
I										
J										

Aircraft Carrier A A A A A

Frigate F F F F

Cruiser C C C

Destroyers D D D D

Submarines S S S

Opponent

	1	2	3	4	5	6	7	8	9	10
A										
B										
C										
D										
E										
F										
G										
H										
I										
J										

Aircraft Carrier A A A A A

Frigate F F F F

Cruiser C C C

Destroyers D D D D

Submarines S S S

Me

	1	2	3	4	5	6	7	8	9	10
A										
B										
C										
D										
E										
F										
G										
H										
I										
J										

Aircraft Carrier A A A A A

Frigate F F F F

Cruiser C C C

Destroyers D D D D

Submarines S S S

Opponent

	1	2	3	4	5	6	7	8	9	10
A										
B										
C										
D										
E										
F										
G										
H										
I										
J										

Aircraft Carrier A A A A A

Frigate F F F F

Cruiser C C C

Destroyers D D D D

Submarines S S S

Me

	1	2	3	4	5	6	7	8	9	10
A										
B										
C										
D										
E										
F										
G										
H										
I										
J										

Aircraft Carrier A A A A A

Frigate F F F F

Cruiser C C C

Destroyers D D D D

Submarines S S S

Opponent

	1	2	3	4	5	6	7	8	9	10
A										
B										
C										
D										
E										
F										
G										
H										
I										
J										

Aircraft Carrier A A A A A

Frigate F F F F

Cruiser C C C

Destroyers D D D D

Submarines S S S

Me

	1	2	3	4	5	6	7	8	9	10
A										
B										
C										
D										
E										
F										
G										
H										
I										
J										

Aircraft Carrier: A A A A A

Frigate: F F F F

Cruiser: C C C

Destroyers: D D D D

Submarines: S S S

Opponent

	1	2	3	4	5	6	7	8	9	10
A										
B										
C										
D										
E										
F										
G										
H										
I										
J										

Aircraft Carrier: A A A A A

Frigate: F F F F

Cruiser: C C C

Destroyers: D D D D

Submarines: S S S

Me

	1	2	3	4	5	6	7	8	9	10
A										
B										
C										
D										
E										
F										
G										
H										
I										
J										

Aircraft Carrier: A A A A A

Frigate: F F F F

Cruiser: C C C

Destroyers: D D D D

Submarines: S S S

Opponent

	1	2	3	4	5	6	7	8	9	10
A										
B										
C										
D										
E										
F										
G										
H										
I										
J										

Aircraft Carrier: A A A A A

Frigate: F F F F

Cruiser: C C C

Destroyers: D D D D

Submarines: S S S

Me

	1	2	3	4	5	6	7	8	9	10
A										
B										
C										
D										
E										
F										
G										
H										
I										
J										

Aircraft Carrier A A A A A

Frigate F F F F

Cruiser C C C

Destroyers D D D D

Submarines S S S

Opponent

	1	2	3	4	5	6	7	8	9	10
A										
B										
C										
D										
E										
F										
G										
H										
I										
J										

Aircraft Carrier A A A A A

Frigate F F F F

Cruiser C C C

Destroyers D D D D

Submarines S S S

Me

	1	2	3	4	5	6	7	8	9	10
A										
B										
C										
D										
E										
F										
G										
H										
I										
J										

Aircraft Carrier: A A A A A

Frigate: F F F F

Cruiser: C C C

Destroyers: D D D D

Submarines: S S S

Opponent

	1	2	3	4	5	6	7	8	9	10
A										
B										
C										
D										
E										
F										
G										
H										
I										
J										

Aircraft Carrier: A A A A A

Frigate: F F F F

Cruiser: C C C

Destroyers: D D D D

Submarines: S S S

Me

	1	2	3	4	5	6	7	8	9	10
A										
B										
C										
D										
E										
F										
G										
H										
I										
J										

Aircraft Carrier A A A A A

Frigate F F F F

Cruiser C C C

Destroyers D D D D

Submarines S S S

Opponent

	1	2	3	4	5	6	7	8	9	10
A										
B										
C										
D										
E										
F										
G										
H										
I										
J										

Aircraft Carrier A A A A A

Frigate F F F F

Cruiser C C C

Destroyers D D D D

Submarines S S S

HOW TO PLAY DOTS AND BOXES

- This is a game for two players.

- Players take it in turns to connect two vertically or horizontally adjacent dots with a pencil. Diagonal lines cannot be made.

- The goal is to own as many squares as possible on the grid. A square is claimed by the player who completes the fourth edge.

- When a player completes a square, they should mark it with their initial.

- Play ends when all the dots are connected.

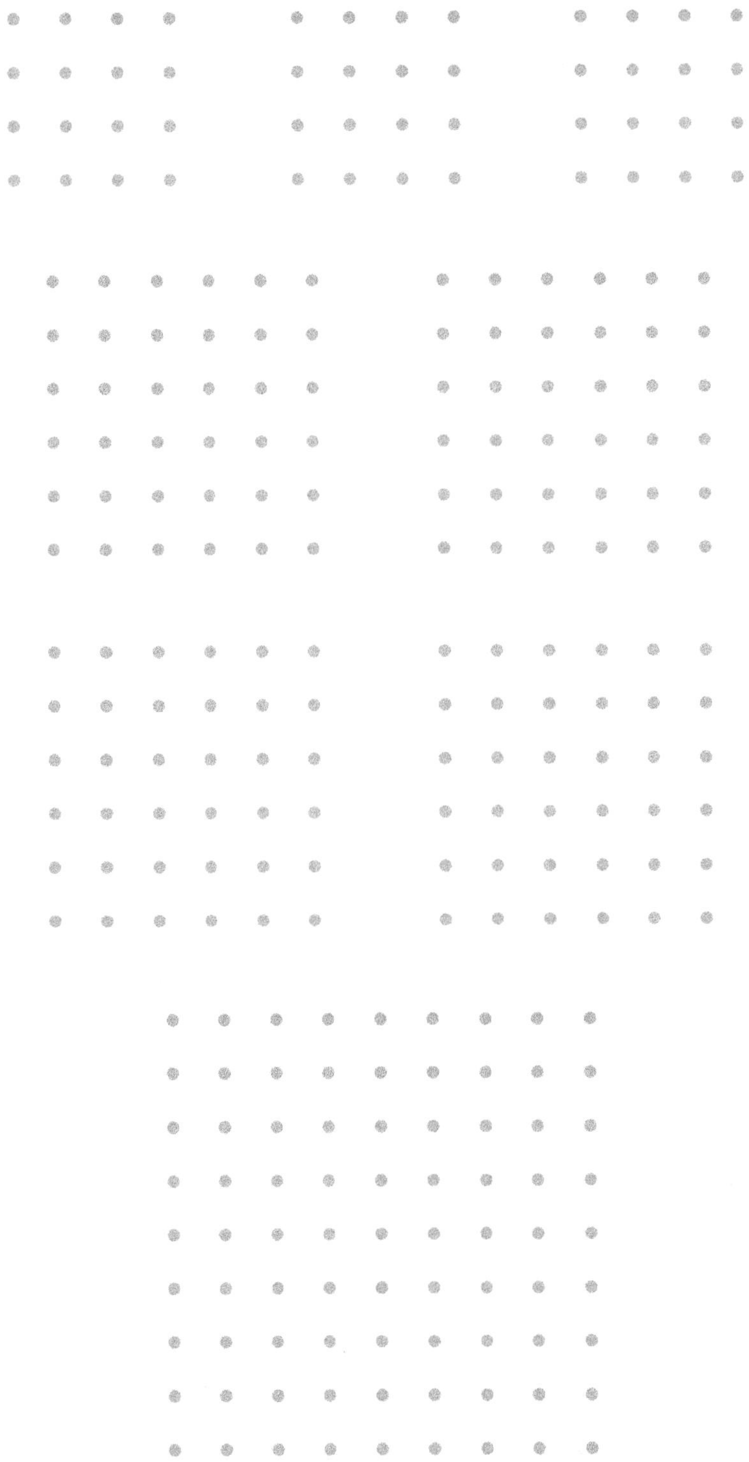

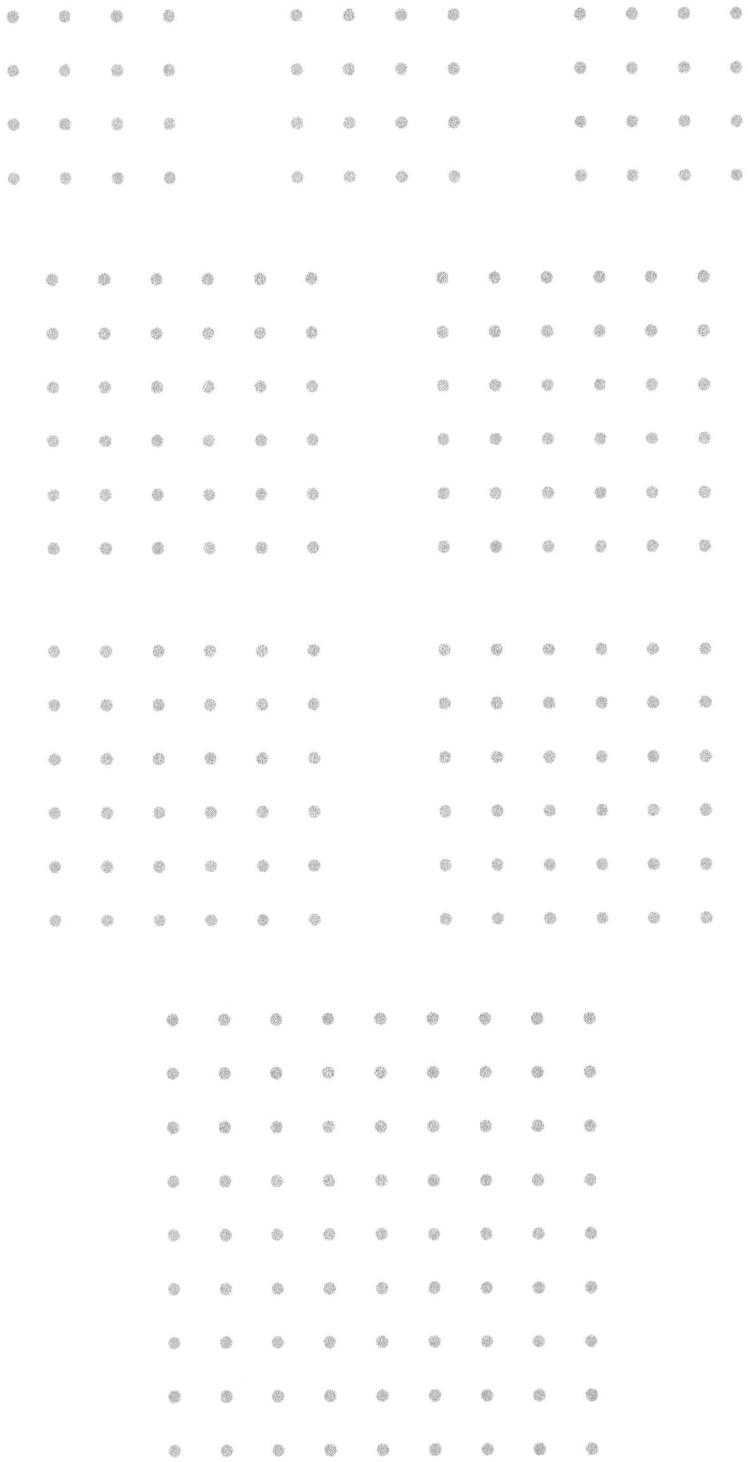

HOW TO PLAY FOUR IN A ROW

Instructions for no gravity variation

• This is a game for two players. They share a board.

• Each player has a different color pen.

• Players take it in turns to color any circle in their color.

• The winner is the first person to connect four circles horizontally, vertically or diagonally.

Instructions for gravity variation

• This is a game for two players. They share a board.

• Each player has a different color pen.

• Players take it in turns to color a circle in their color. They may only color the bottom-most empty circle in any column.

• The winner is the first person to connect four circles horizontally, vertically or diagonally.

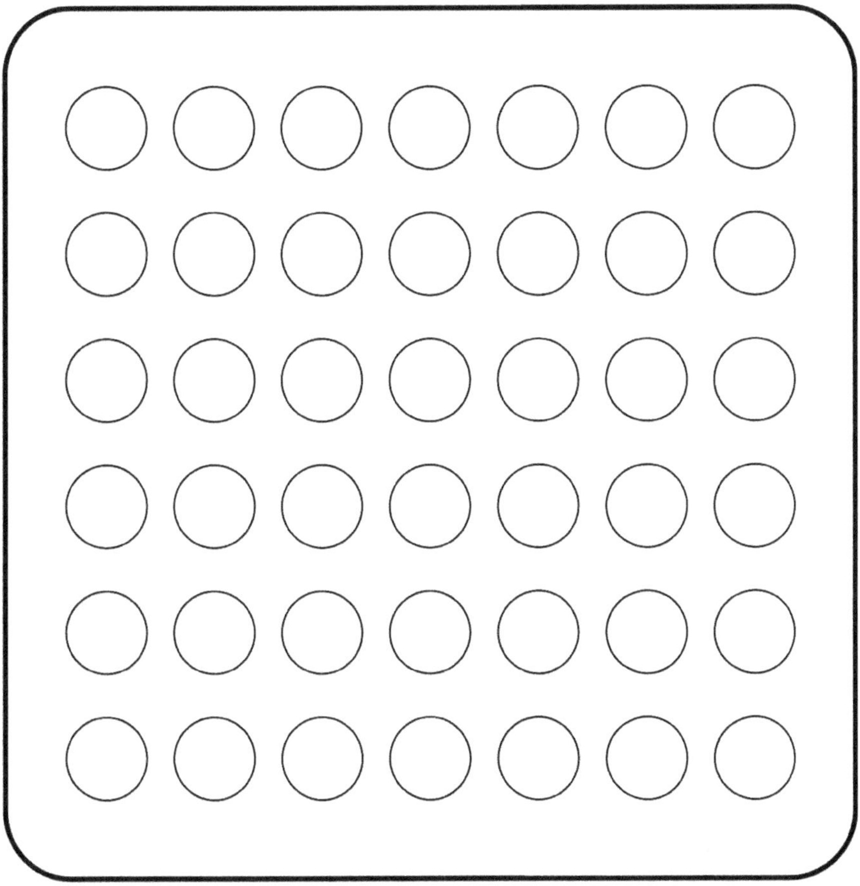

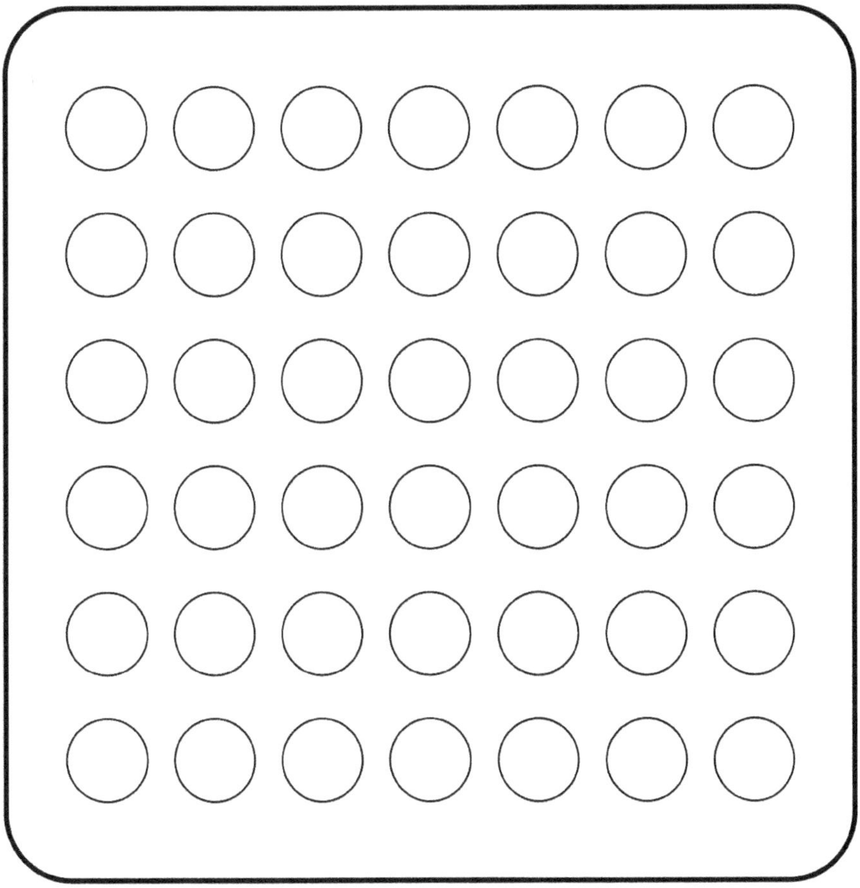

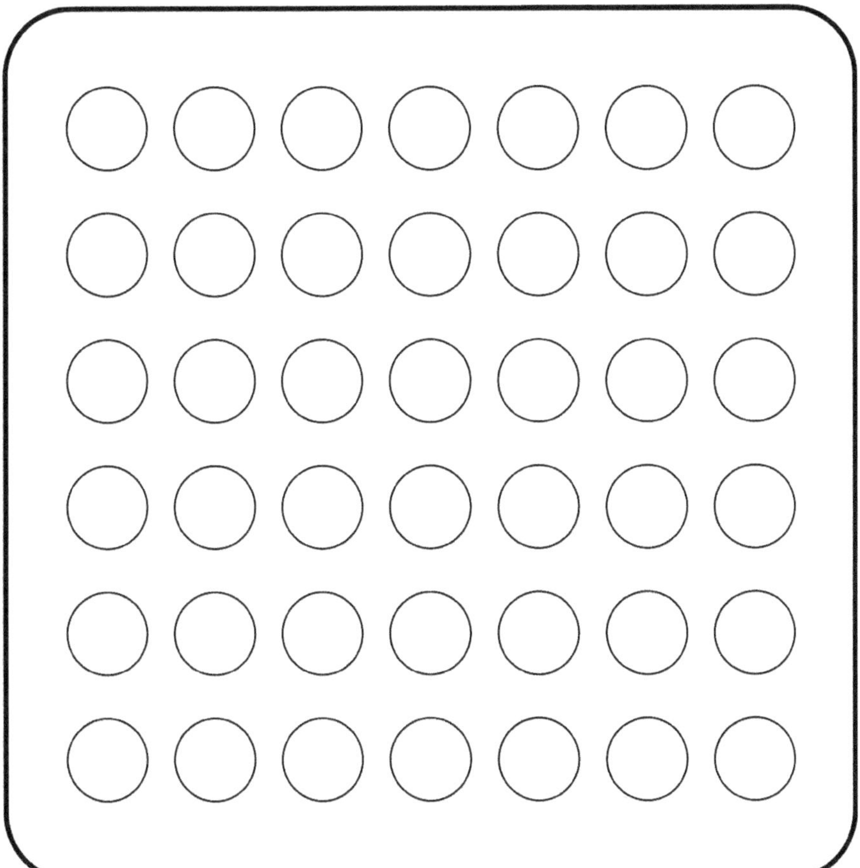

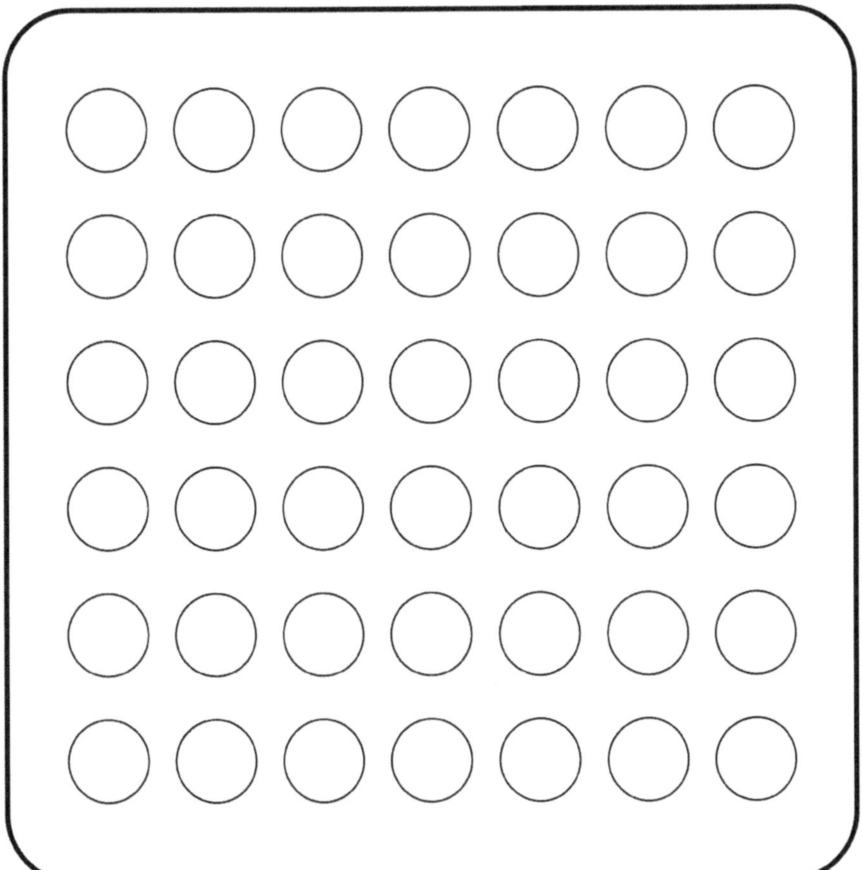

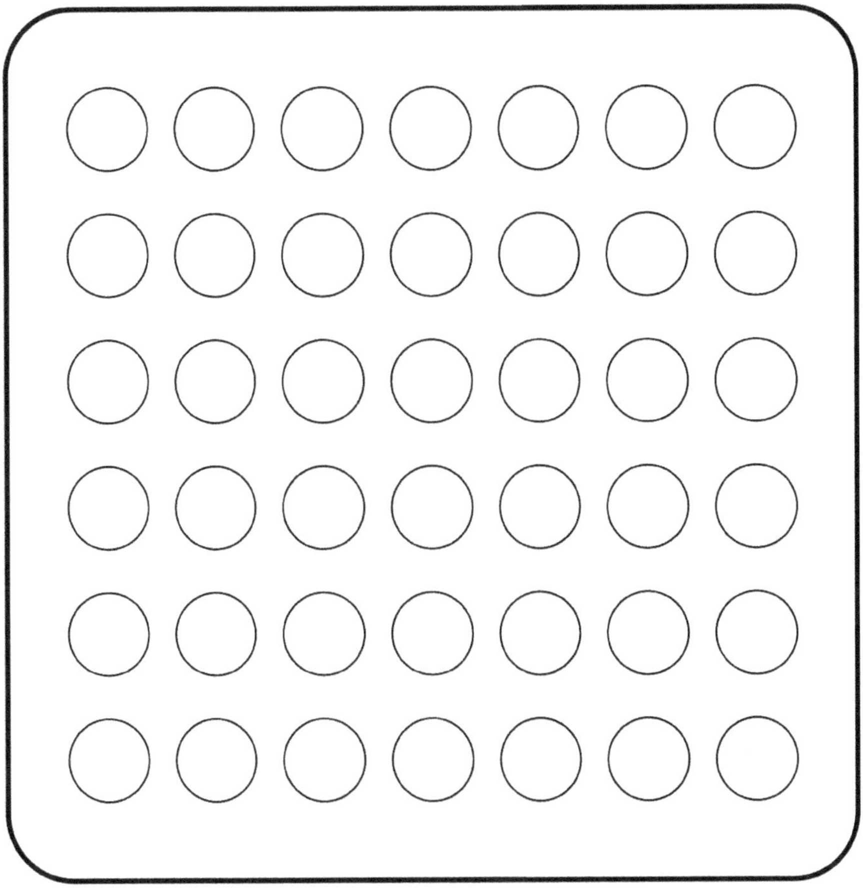

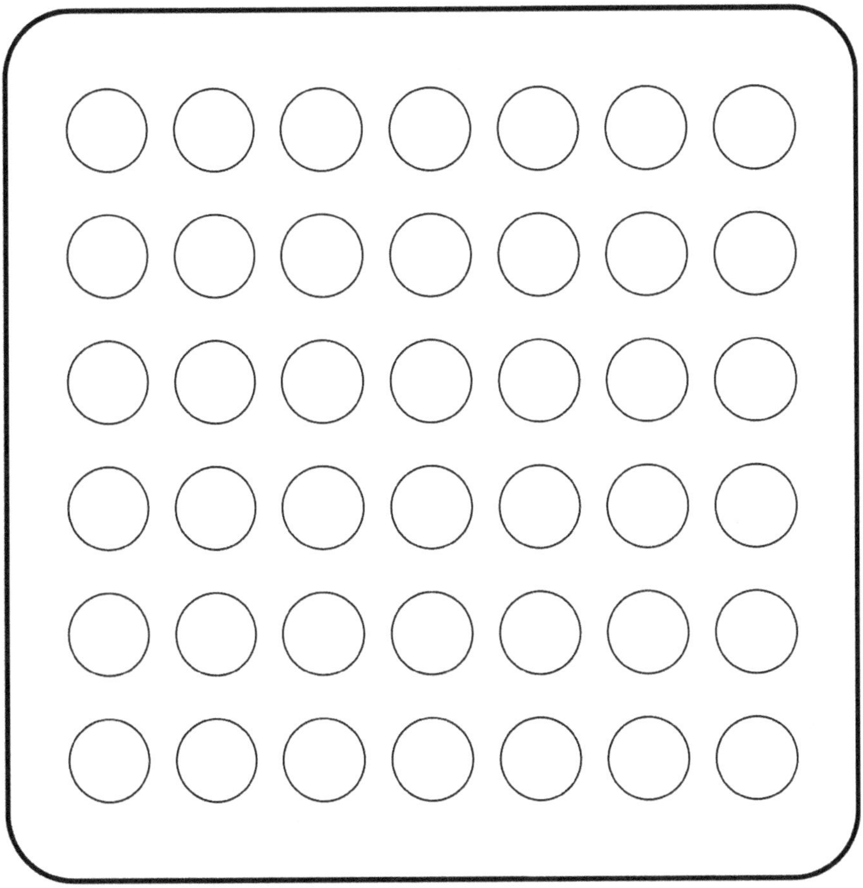

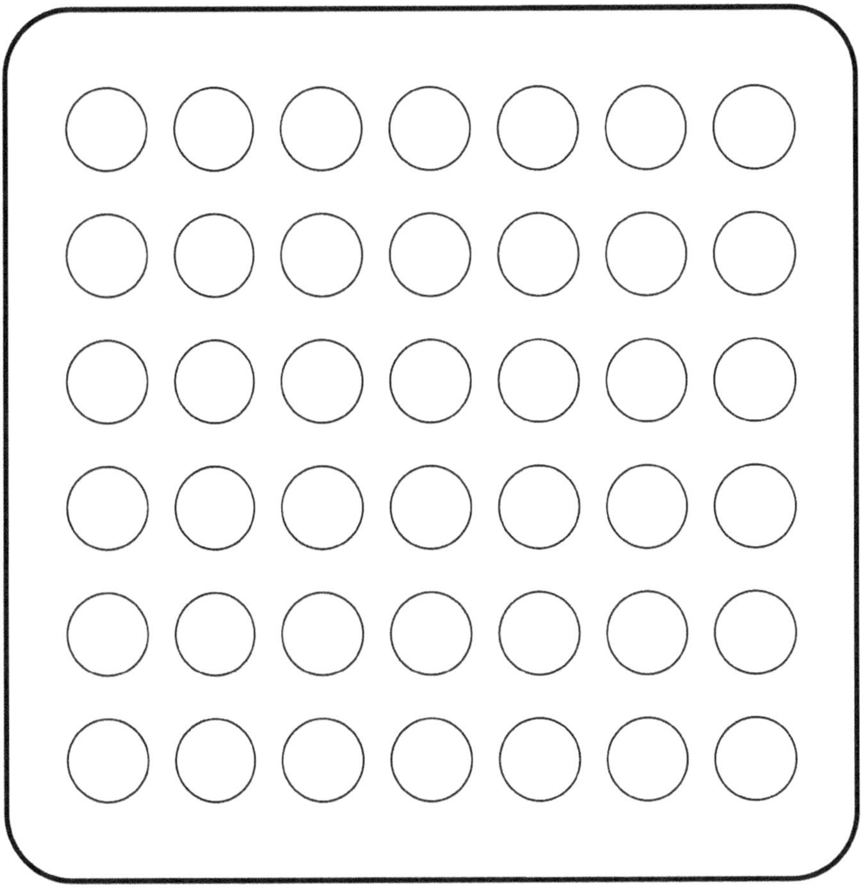

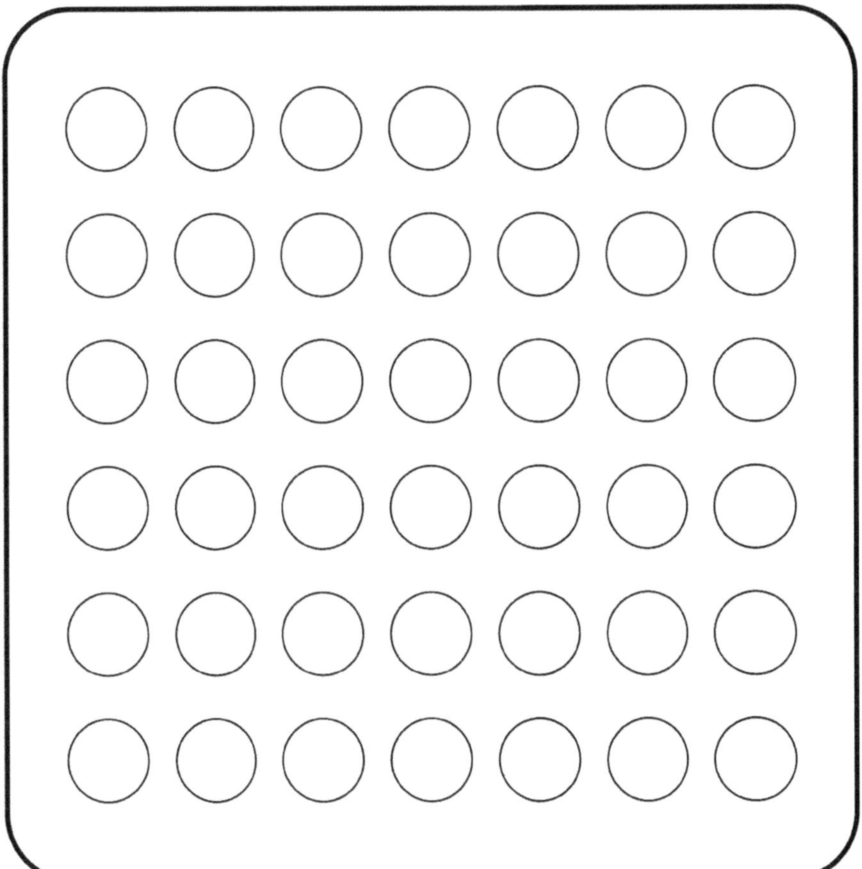

HOW TO PLAY HEXAGON GAME

- This is a game for two players.

- Each player should select a different colored pen or pencil (traditionally blue and red)

- Players take it in turns to join two dots.

- The goal is to avoid completing a triangle with all three sides in your own color.

- The first person to complete a triangle in their color loses.

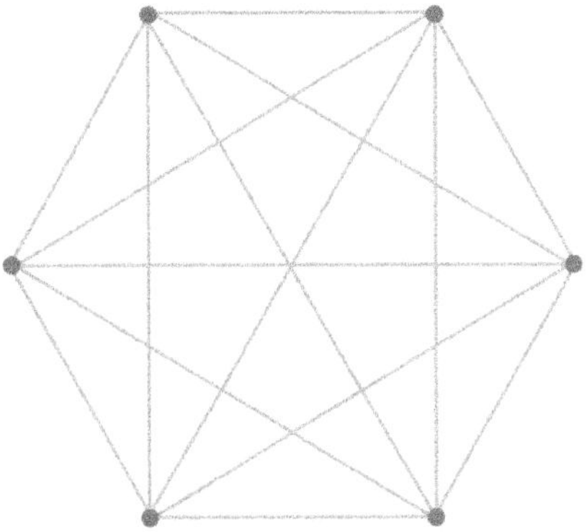

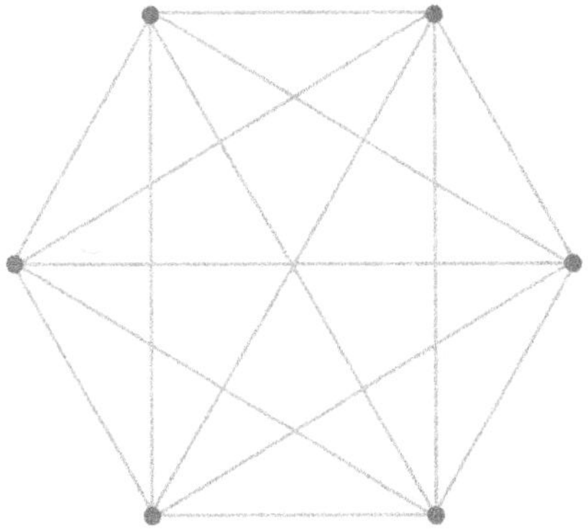

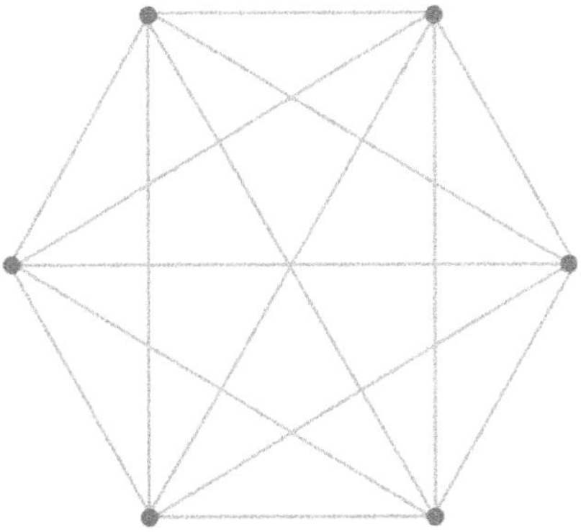

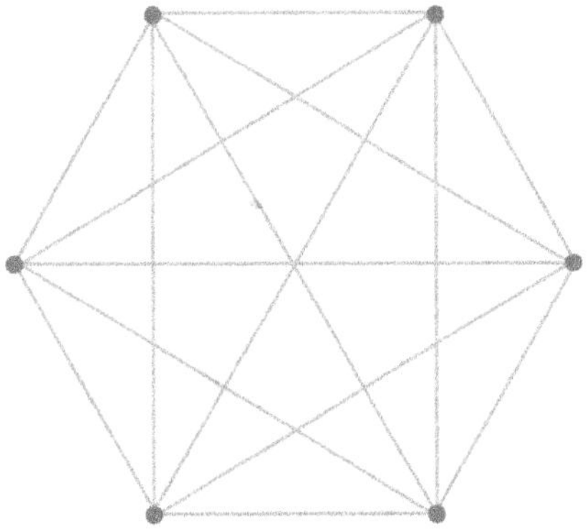

Player 1

Player 2

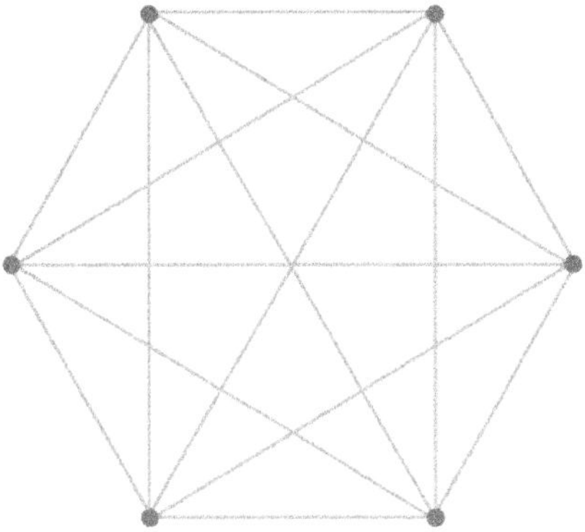

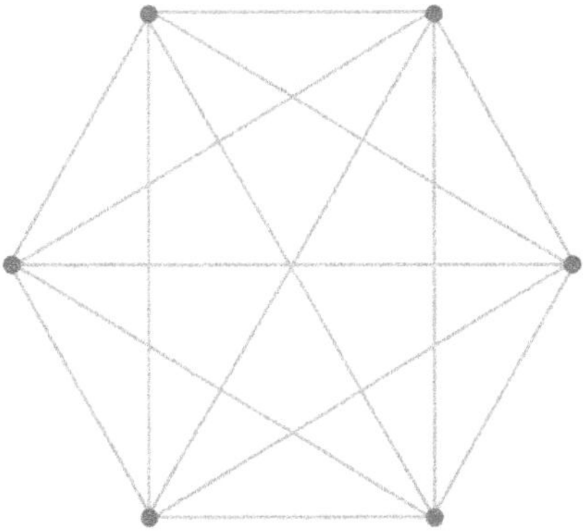

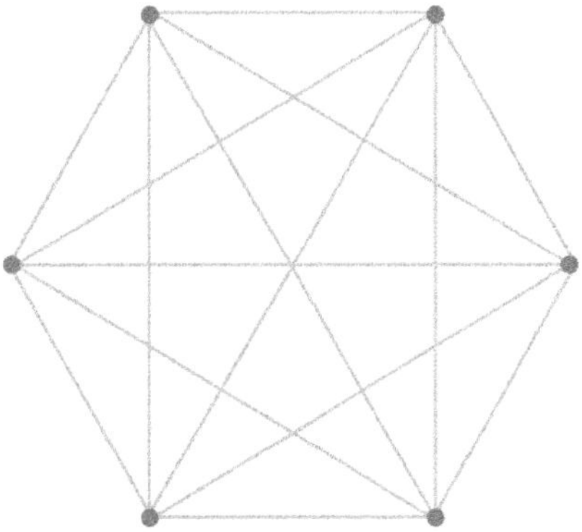

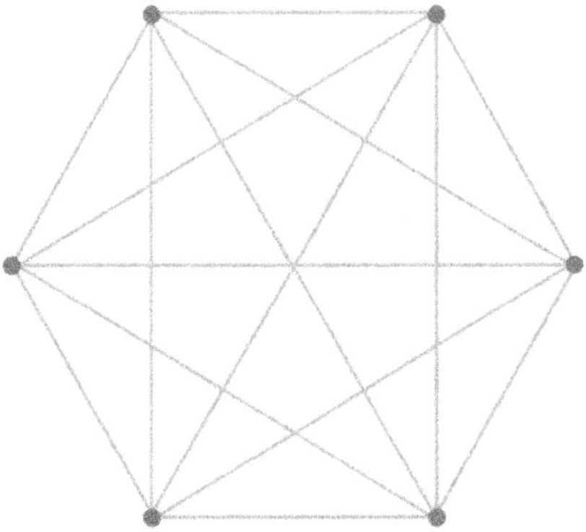

Player 1 **Player 2**

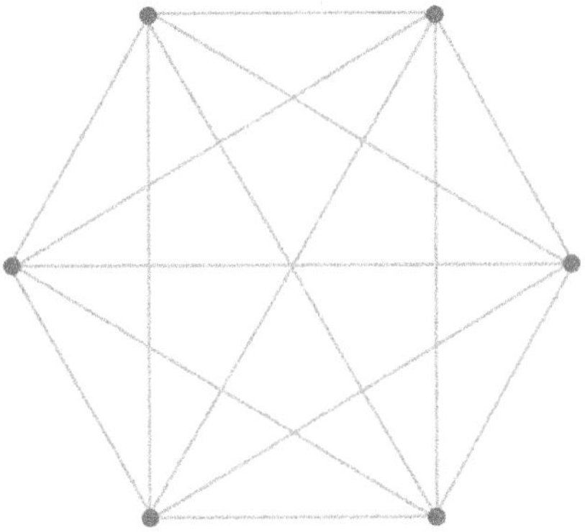

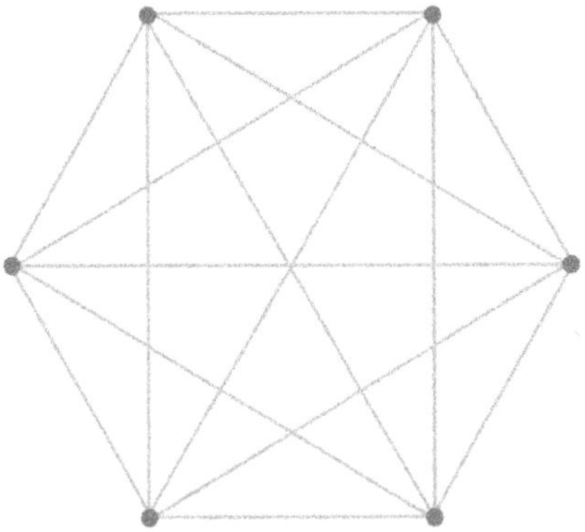

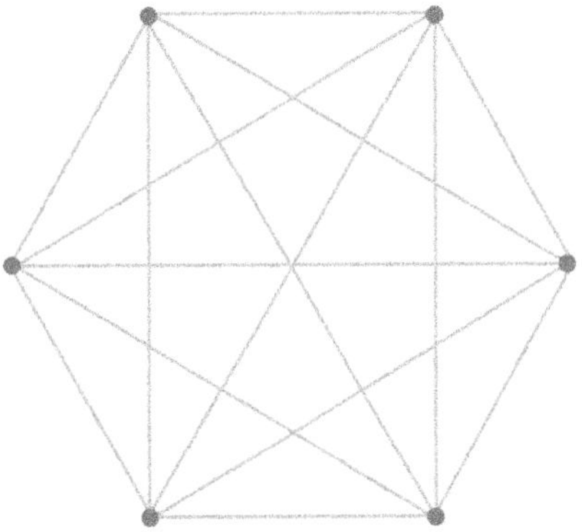

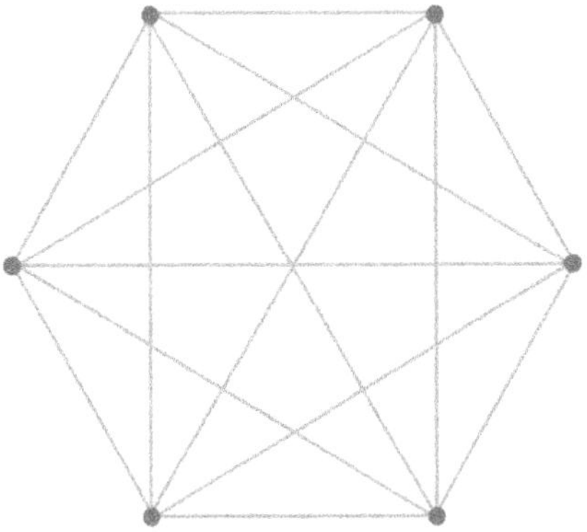

Player 1 **Player 2**

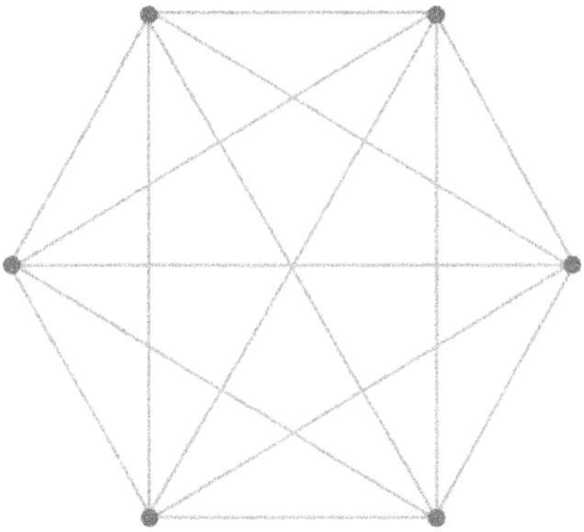

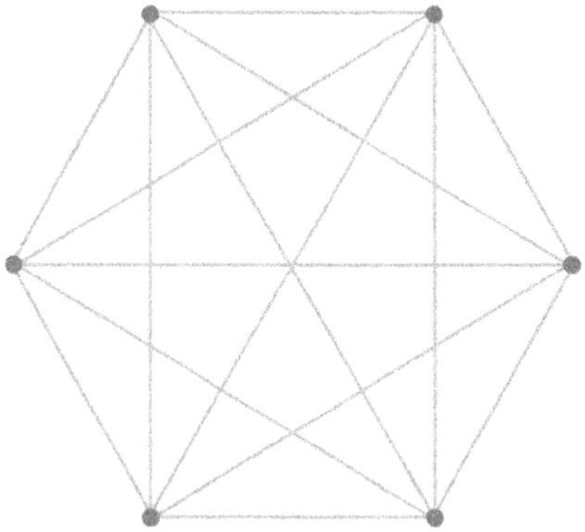

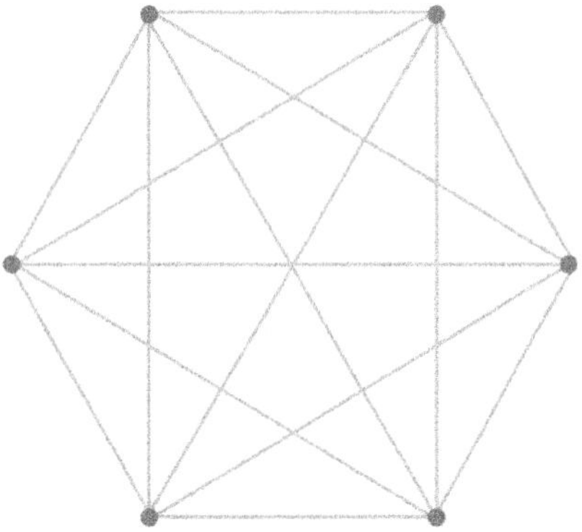

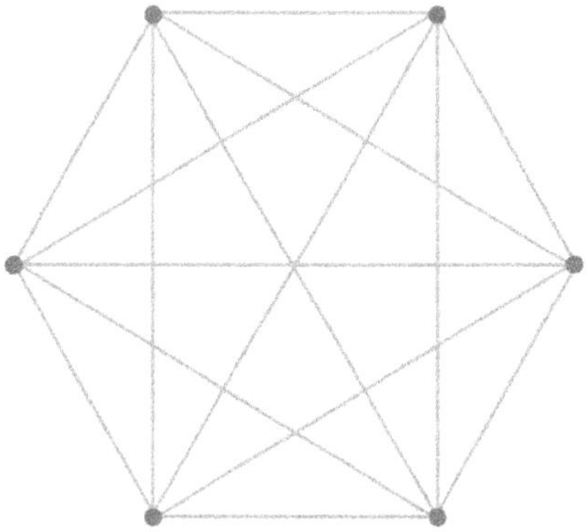

Player 1

Player 2

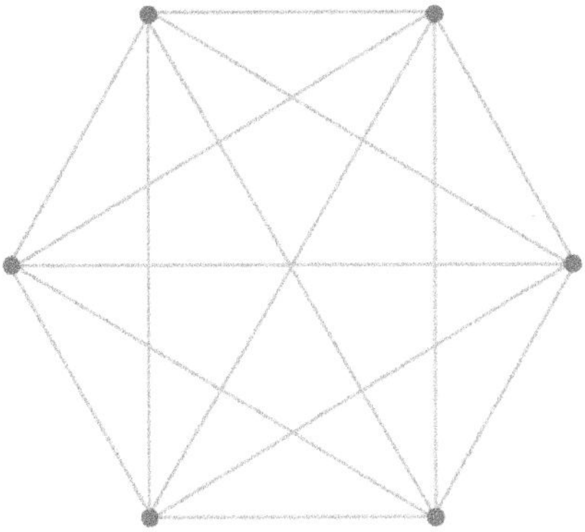

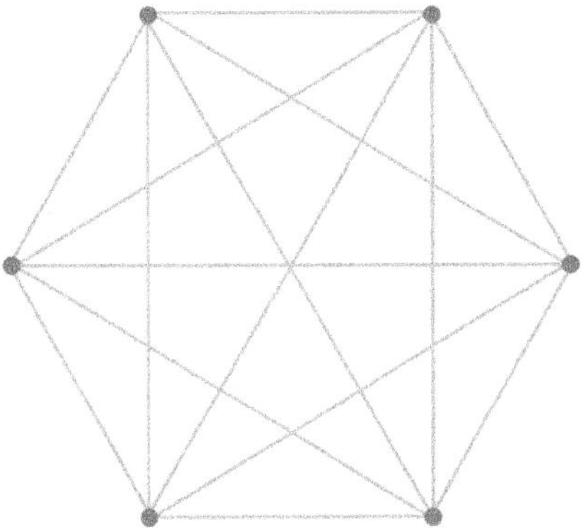

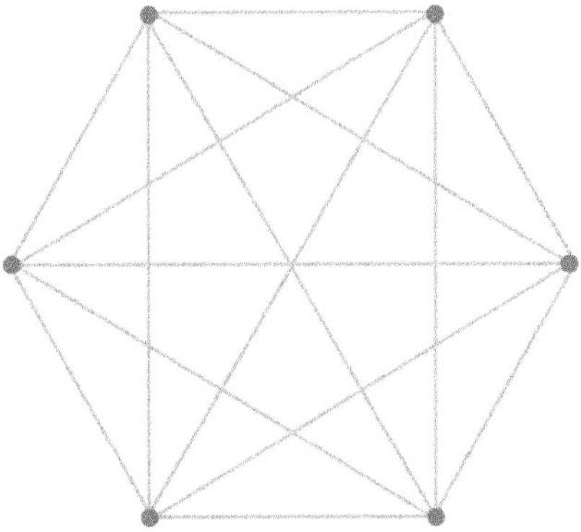

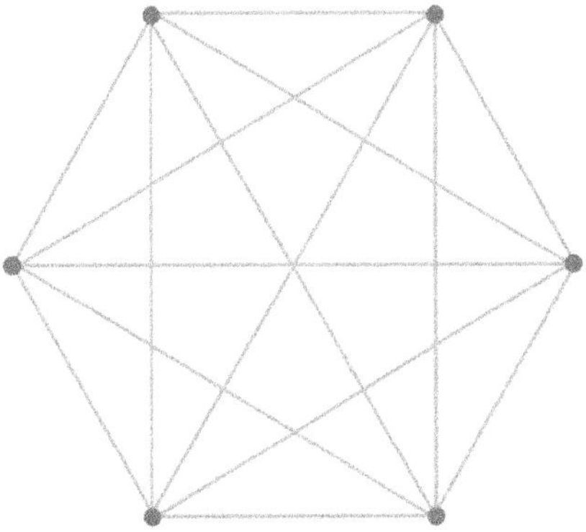

Player 1

Player 2

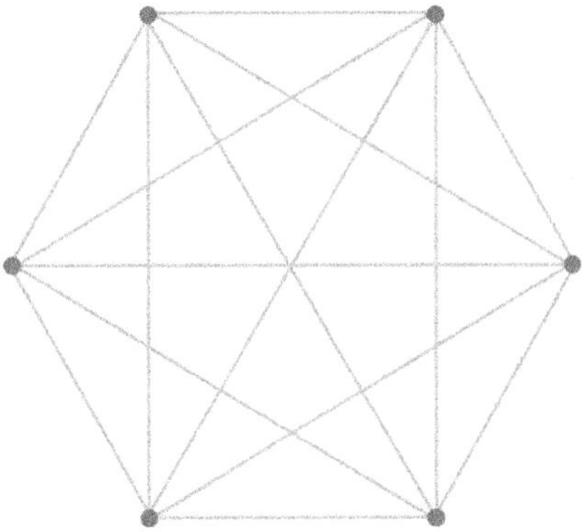

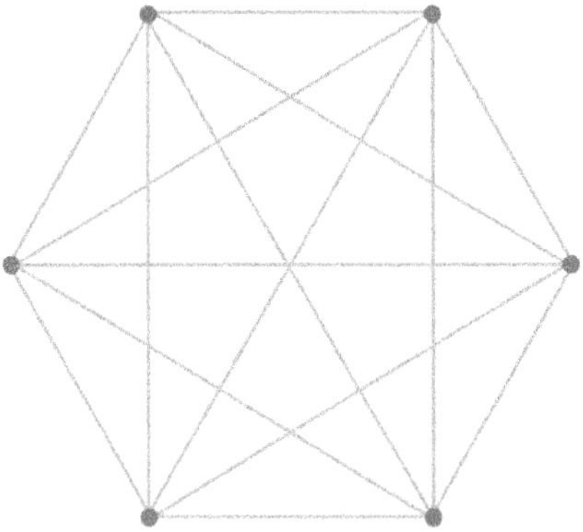

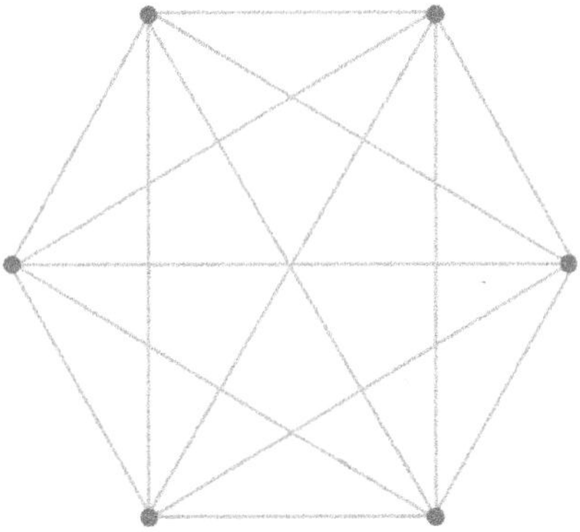

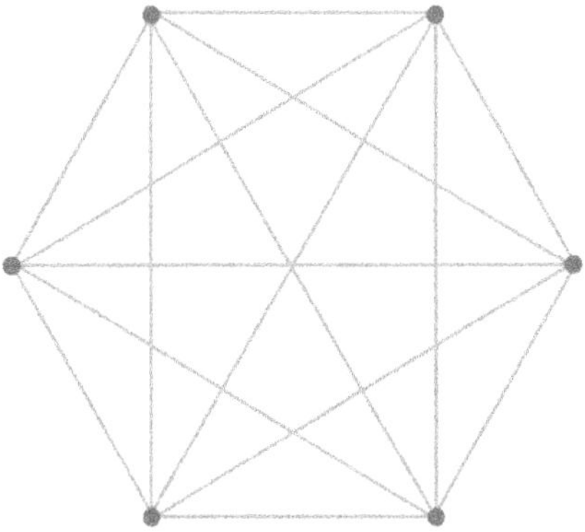

Player 1

Player 2

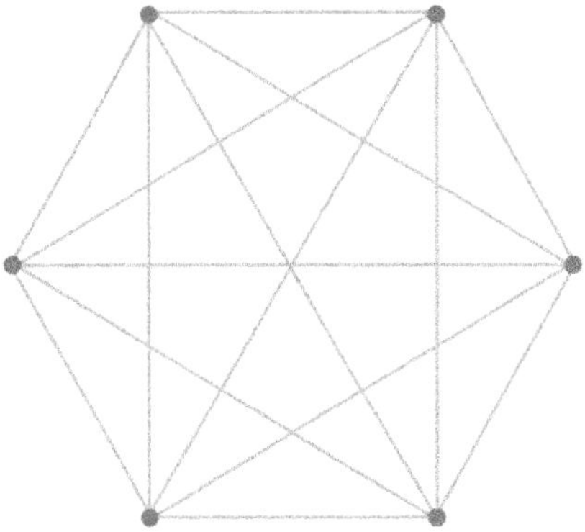

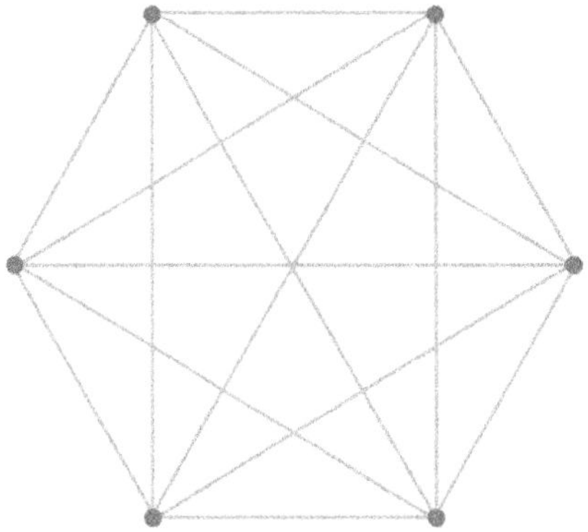

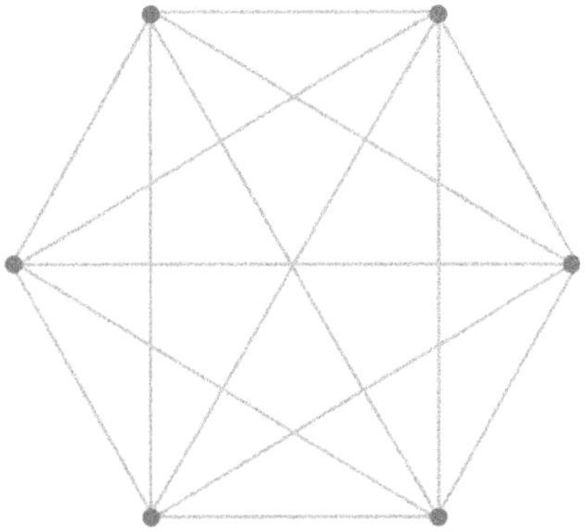

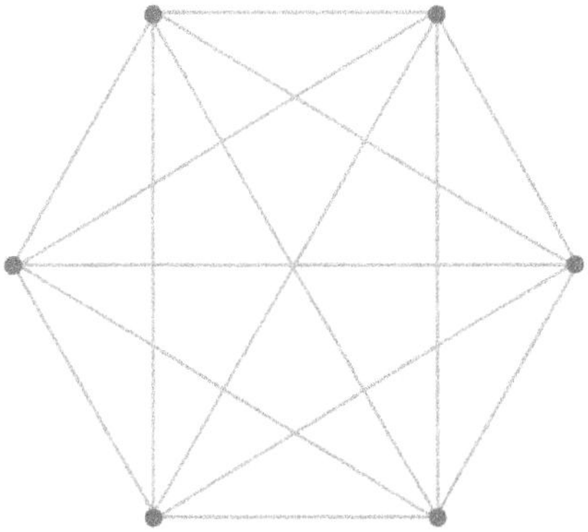

Player 1

Player 2

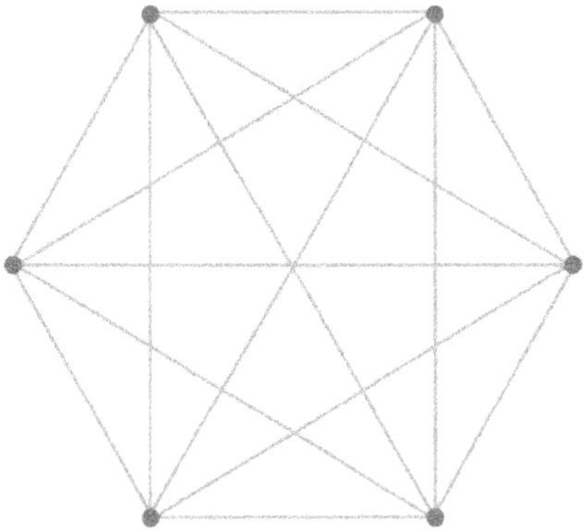

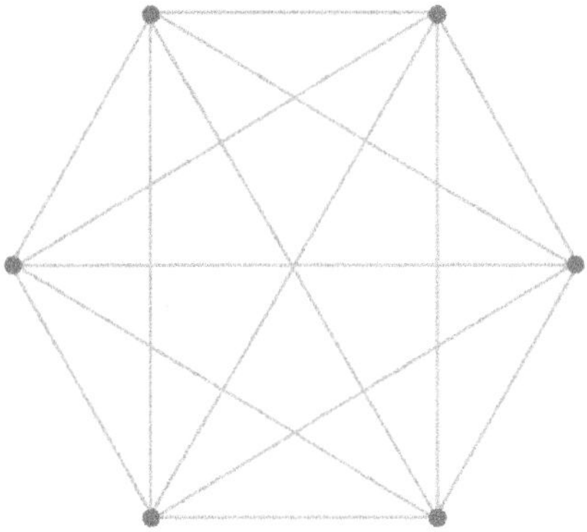

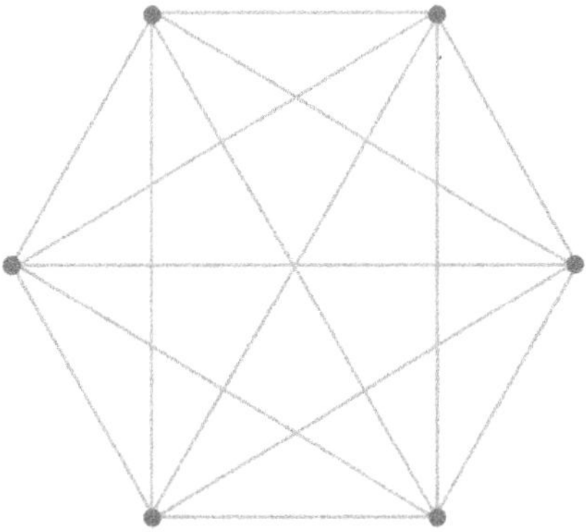

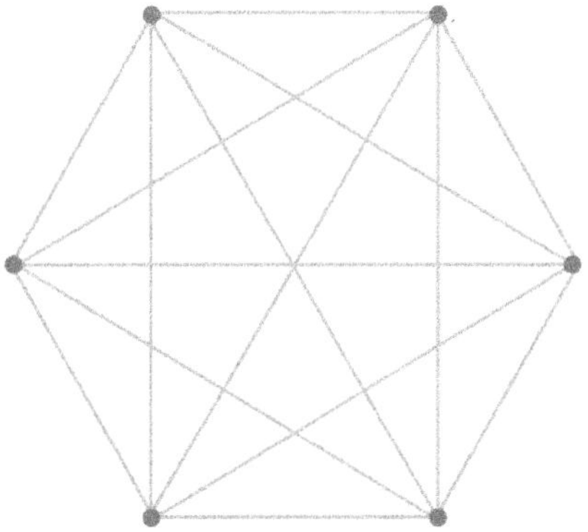

Player 1 ⬡ **Player 2** ⬡

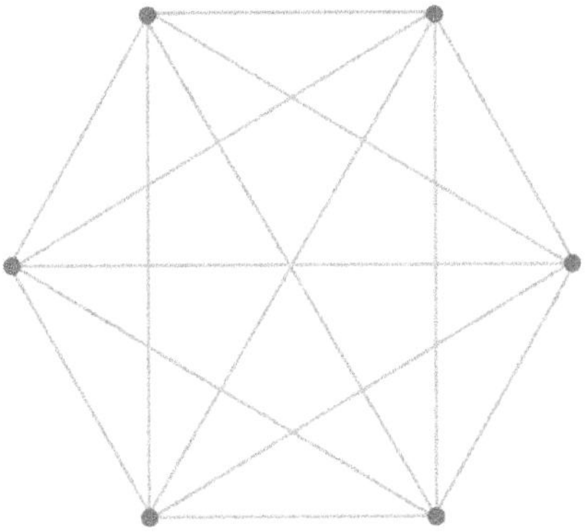

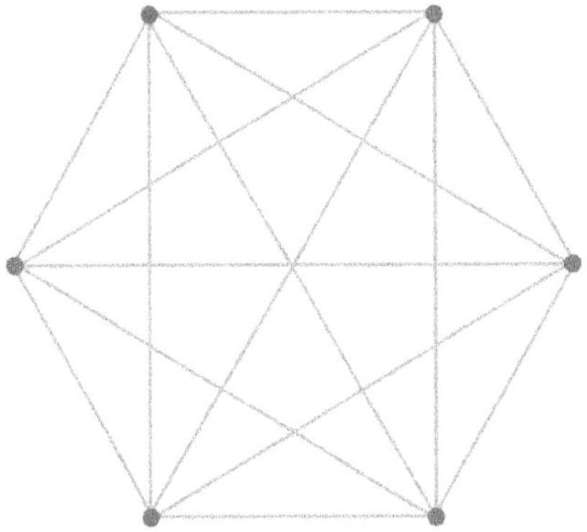

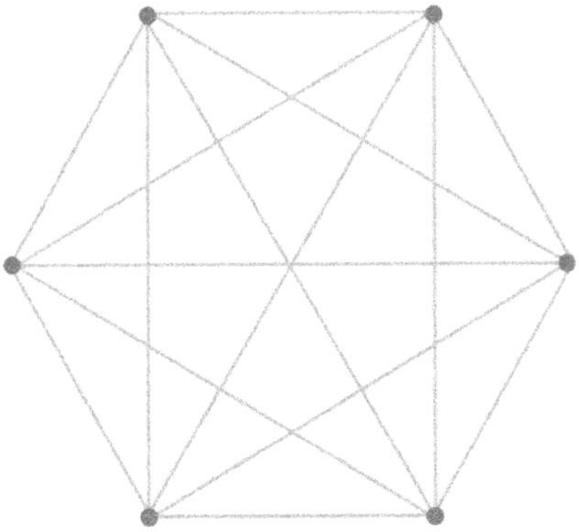

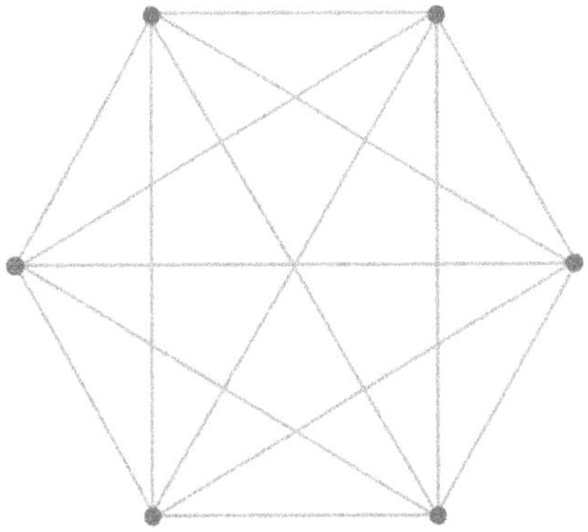

Player 1 **Player 2**

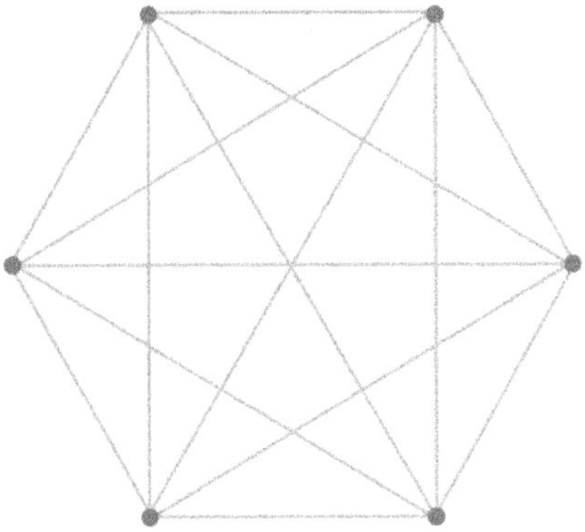

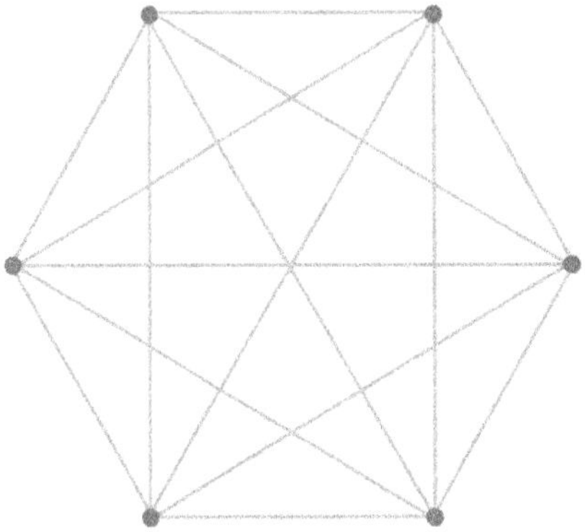

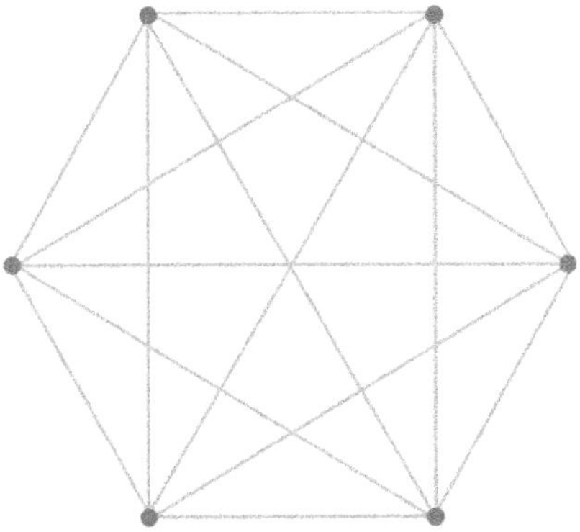

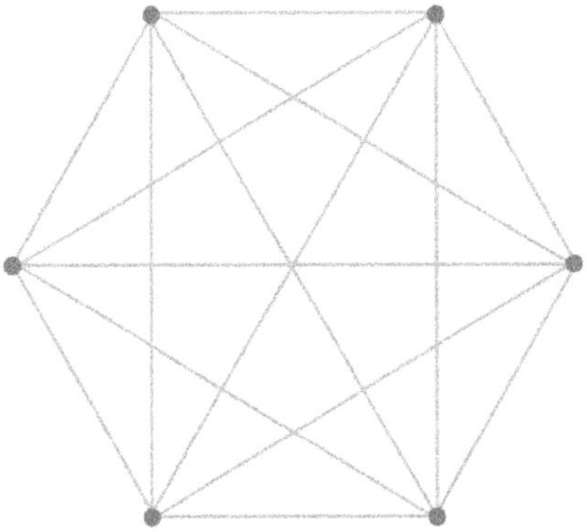

Player 1 **Player 2**

www.ingramcontent.com/pod-product-compliance
Lightning Source LLC
Chambersburg PA
CBHW071421210526
45465CB00001B/475
* 9 7 9 8 6 4 4 0 2 7 9 1 0 *